BRINGING BACK THE California Condor

Rachel Stuckey

CRABTREE
PUBLISHING COMPANY
WWW.CRABTREEBOOKS.COM

Author: Rachel Stuckey

Series Research and Development: Reagan Miller

Picture Manager: Sophie Mortimer

Design Manager: Keith Davis

Editorial Director: Lindsey Lowe

Children's Publisher: Anne O'Daly

Editor: Ellen Rodger

Proofreader: Crystal Sikkens

Cover design: Margaret Amy Salter

Production coordinator and Prepress technician: Margaret Amy Salter

Print coordinator: Katherine Berti

Produced for Crabtree Publishing Company
by Brown Bear Books

Photographs
(t=top, b= bottom, l=left, r=right, c=center)

Front Cover: All images from Shutterstock

Interior: Alamy: Minden Pictures 14, 17t, 17b, 24, Natural History Collection 29, Chuck Pace 13, Maurice Savage 15b, USFWS Photo 23; Dreamstime: Michael Elliot 15t, Scott Jones 9t; Getty Images: Los Angeles Times/Mel Melcon 20; iStock: JPL Designs 8, Chris Fairfax 12, J Kauffeld 7, LICreate 19, Nick Lust Photography 5, On Location 1, Png Studio 16, W. Antoinette 27b; Kain Road Cul de Sac: 9b; Dawn Kish: 18; Shutterstock: Agami Photo Agency 27t, Barabara Ash 4, EcoPrint 26, Necha Evkon 25, Patrick Jennings 22, Brian A Wolf 21, Zenobillis 6; U.S. Fish and Wildlife Service: 28, Joseph Brandt 10.

Brown Bear Books has made every attempt to contact the copyright holder. If you have any information please contact licensing@ brownbearbooks.co.uk

Library and Archives Canada Cataloguing in Publication

Title: Bringing back the California condor / Rachel Stuckey.
Names: Stuckey, Rachel, author.
Series: Animals back from the brink.
Description: Series statement: Animals back from the brink | Includes index.
Identifiers: Canadiana (print) 20190233370 | Canadiana (ebook) 20190233389 | ISBN 9780778768111 (hardcover) | ISBN 9780778768258 (softcover) | ISBN 9781427124265 (HTML)
Subjects: LCSH: California condor—North America—Juvenile literature. | LCSH: California condor—Conservation—North America—Juvenile literature. | LCSH: Endangered species—North America—Juvenile literature. | 0LCSH: Wildlife recovery—North America—Juvenile literature.
Classification: LCC QL696.C53 S78 2020 | DDC j598.9/2097—dc23

Library of Congress Cataloging-in-Publication Data

Names: Stuckey, Rachel, author.
Title: Bringing back the California condor / Rachel Stuckey.
Other titles: Animals back from the brink.
Description: New York, New York : Crabtree Publishing Company, [2020] | Series: Animals back from the brink | Includes index.
Identifiers: LCCN 2019053201 (print) | LCCN 2019053202 (ebook) | ISBN 9780778768111 (hardcover) | ISBN 9780778768258 (paperback) | ISBN 9781427124265 (ebook)
Subjects: LCSH: California condor--Conservation--Juvenile literature. | Wildlife conservation--Juvenile literature.
Classification: LCC QL696.C53 S78 2020 (print) | LCC QL696.C53 (ebook) | DDC 598.9/2--dc23
LC record available at https://lccn.loc.gov/2019053201
LC ebook record available at https://lccn.loc.gov/2019053202

Crabtree Publishing Company
www.crabtreebooks.com 1-800-387-7650

Printed in Canada/022023/CPC20230217

Published in Canada
Crabtree Publishing
616 Welland Ave.
St. Catharines, Ontario
L2M 5V6

Published in the United States
Crabtree Publishing
347 Fifth Ave
Suite 1402-145
New York, NY 10016

Published in the United Kingdom
Crabtree Publishing
Maritime House
Basin Road North, Hove
BN41 1WR

Published in Australia
Crabtree Publishing
Unit 3-5
Currumbin Court
Capalaba QLD 4157

Contents

Snapshot of the Condor

California condors are **New World vultures**. These are a different **species** from the vultures found in Asia, Africa, and Europe. California condors are among the longest-living birds in the world, surviving for up to 60 years. Until the 1700s, there were thousands of California condors in the wild. However, during the expansion of U.S. settlement in the 1800s, hunting and **habitat** destruction led to a steady decline in bird numbers. By 1967, the California condor became one of the first species to be protected under the new Endangered Species Preservation Act. Despite this, by the 1980s there were only 22 wild birds left in southern California. In 1987, it was listed as **Extinct** in the Wild (EW).

The California condor is the largest flying land bird in North America, with a wingspan of 9.8 feet (3 m). The bird flaps its wings to take flight, but then soars using wind currents to stay in the air. They can travel at up to 56 miles per hour (90 kph) and reach heights of up to 15,000 feet (4,600 m), which is higher than the Rocky Mountains.

FAST FOOD

Condors have black feathers with white patches on the underside of their wings. They have bald heads with skin colors in shades of gray, pink, red, yellow or bright orange. They have long, powerful beaks, but unlike hawks and eagles, vultures do not hunt and kill food. Condors are **scavengers** that eat **carrion**, or the flesh of animals that have already died, such as deer, cattle, and sheep. Because they have no sense of smell, they use their good eyesight to find food. When they spot a **carcass** from the air, they circle and swoop down to start picking at it. Sometimes it looks as if condors are fighting over carrion, but often two birds will rock back and forth if the animal is big, using their combined body weight to tear the meat apart.

Today, over 150 California condors live in the wild in California. A large population lives in the Pinnacles National Park, where they soar thousands of feet into the sky on thermals, or rising spirals, of air. This bird (circled in red) was seen at moonrise.

Species at Risk

Created in 1984, the International Union for the **Conservation** of Nature (IUCN) protects wildlife, plants, and natural resources around the world. Its members include about 1,400 governments and nongovernmental organizations. The IUCN publishes the Red List of Threatened Species each year, which tells people how likely a plant or animal species is to become extinct. It began publishing the list in 1964.

In 2018, Spix's Macaw from Brazil was listed as Critically Endangered (CR), but probably is now Extinct in the Wild (EW). The last known wild bird has not been seen since 2000. The IUCN updates its Red List twice a year and reevaluates each species every five years.

SCIENTIFIC CRITERIA

The Red List, created by scientists, divides nearly 80,000 species of plants and animals into nine categories. Criteria for each category include the growth and decline of the population size of a species. They also include how many individuals within a species can breed, or have babies. In addition, scientists include information about the habitat of the species, such as its size and quality. These criteria allow scientists to figure out the probability of extinction facing the species.

IUCN LEVELS OF THREAT

The Red List uses nine categories to define the threat to a species.

Extinct (EX)	No living individuals survive.
Extinct in the Wild (EW)	Species cannot be found in its natural habitat. Exists only in captivity, in cultivation, or in an area that is not its natural habitat.
Critically Endangered (CR)	At extremely high risk of becoming extinct in the wild.
Endangered (EN)	At very high risk of extinction in the wild.
Vulnerable (VU)	At high risk of extinction in the wild.
Near Threatened (NT)	Likely to become threatened in the near future.
Least Concern (LC)	Widespread, abundant, or at low risk.
Data Deficient (DD)	Not enough data to make a judgment about the species.
Not Evaluated (NE)	Not yet evaluated against the criteria.

In the United States, the Endangered Species Act of 1973 was passed to protect species from possible extinction. It has its own criteria for classifying species, but they are similar to those of the IUCN. Canada introduced the Species at Risk Act in 2002. More than 530 species are protected under the act. The list of species is compiled by the Committee on the Status of Endangered Wildlife in Canada (COSEWIC).

CONDORS AT RISK

In 2018, the IUCN Red List classified the California condor as Critically Endangered (CR). In the 1980s, only 22 birds were left in the wild. They were taken into captivity for protection. By 1987, the condor was officially Extinct in the Wild (EW) because all surviving condors were in captivity. Today, there are condors in California, Mexico, Arizona, and Utah. They number around 500, but only 44 were bred in the wild. The species remains at risk of extinction.

The Cleanup Crew

Condors live in rocky shrubland, coniferous forests, and oak savannas. They are often found near cliffs or large trees, which they use as nesting sites. Birds may travel up to 160 miles (250 km) in search of food. The California condor is a scavenger, and all scavengers are an important part of the **ecosystem**. It may seem gross that condors eat dead animals, but condors help to get rid of the decaying bodies before they poison the environment. This helps to prevent the spread of disease in the ecosystem. The bacteria that helps to decompose or break down the carrion tissue helps condors to digest their food easily.

Without important scavengers such as the California condor, the environment would soon be polluted by slowly rotting animal carcasses.

LEAD POISONING

Lead poisoning is the number one cause of death for adult California condors. In the late 1800s, scientists discovered that lead is poisonous if swallowed by humans and other animals. Today, it is illegal to use lead in products like paint, gasoline, and water pipes. However, lead is still used to make bullets for hunting rifles. When an animal is shot by a lead bullet, small particles of lead spread through its body. If a condor feeds on the body, the lead particles break down in the condor's stomach and enter its bloodstream. Over time, the poison builds up and kills the condor.

THE PROBLEM OF "MICROTRASH"

Condors are curious birds attracted by interesting bits of trash. Unfortunately, they eat a lot of microtrash. This is small pieces of trash such as broken glass (right), bottle caps, soda can tabs, and broken down pieces of trash or garbage. Condors feed microtrash to their chicks by **regurgitation**. If it gets stuck in the young condors' stomach, they will not be able to digest food and will starve to death. This is the most common cause of death in young condor chicks.

Population in Decline

Condors nest in large holes in the sides of cliffs or in giant trees such as sequoias and redwoods. As increasing numbers of settlers moved to the western United States and built towns, cities, and roads, the habitat of the California condor began to disappear. It became harder for them to breed, raise their young, and find food. The power lines that carried electricity to homes and businesses also threatened the condors and other birds. Birds collided with the poles and were **electrocuted** when they perched on the wires. By 1967, the California condor population was endangered.

A condor egg hatches after 54 to 58 days. The parents take turns keeping the chick warm and fed. At about six months old, the flying feathers are fully grown. The chick stays with its parents for two years while it learns how to fly and scavenge.

BREEDING PROBLEMS

California condors have a naturally low rate of birth. It takes six years for condors to reach breeding age. After mating, female condors only lay one egg at a time, and **breeding pairs** only raise one **nestling** every two years. Condor eggs are a food source for ravens, and baby condors make easy prey for golden eagles and bears. If a breeding pair of condors lose an egg in the first month, the female may lay another egg. However, if the nestling is lost, she will not lay another egg for two years. This made it hard for condors to increase their falling population. They needed human help.

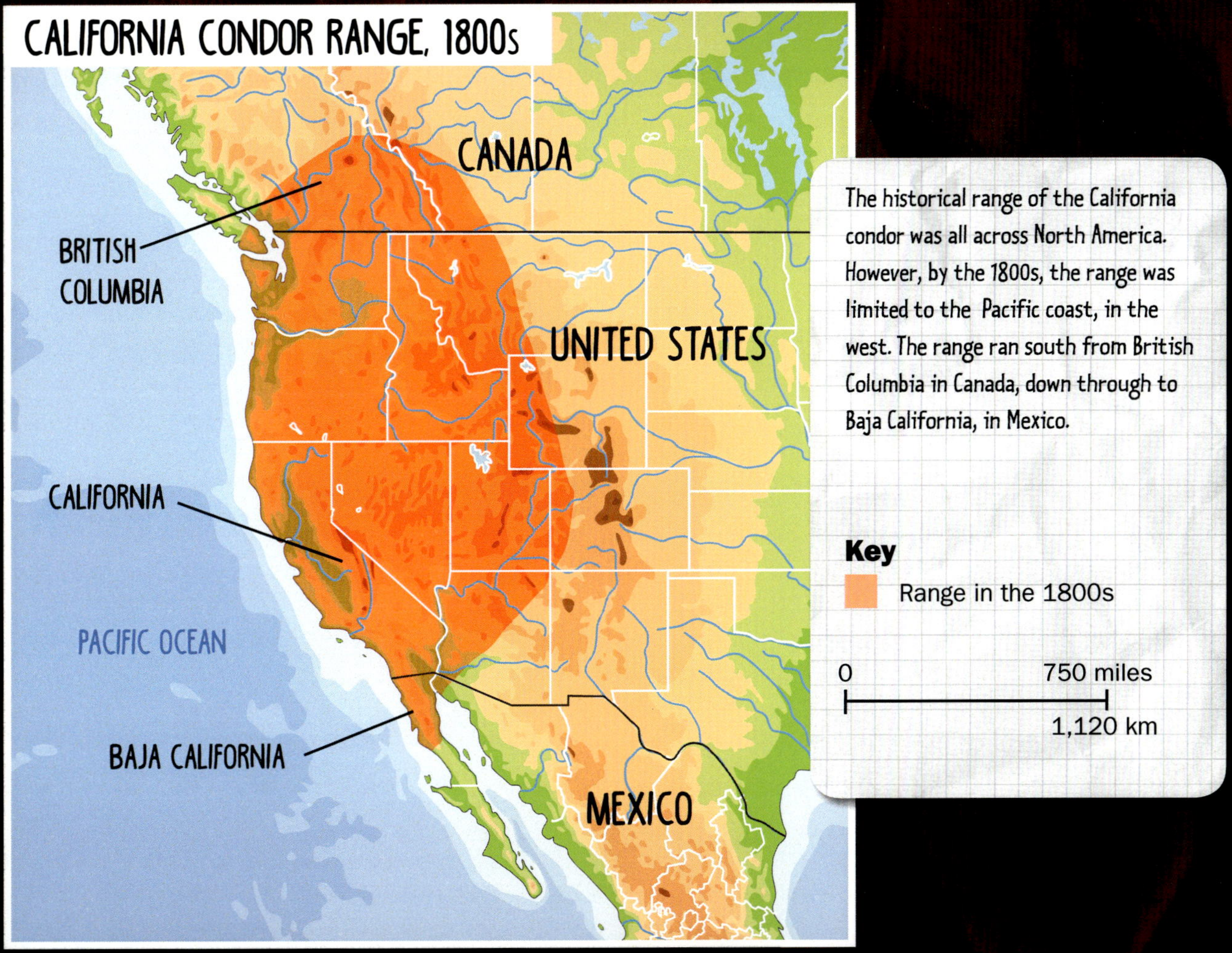

The historical range of the California condor was all across North America. However, by the 1800s, the range was limited to the Pacific coast, in the west. The range ran south from British Columbia in Canada, down through to Baja California, in Mexico.

Who Got Involved?

By 1982, the wild California condor population was down to just 22 birds. The U.S. authorities began a conservation project to save the species. Zoos in Los Angeles, Oakland, and San Diego joined the recovery effort. Conservation organizations such as the Peregrine Fund and the Ventana Wildlife Society also began to work closely with zoos and government agencies. The Peregrine Fund was founded in 1970 to help save peregrine falcons. The Ventana Wildlife Society releases captive condors into the protected Ventana Wilderness of Los Padres National Forest, and the Pinnacles National Park in Salinas, California.

Zoos were given official permission to start the first California Condor Recovery Program by the U.S. Fish & Wildlife Service and the California Department of Fish & Game. Here, a California condor released by the Ventana Wildlife Society lands on a rock in the Pinnacles National Park.

The Santa Barbara Zoo became an official member of the California Condor Recovery Program in 2002. It plays an important role in the reintroduction of captive-bred condors to the wild. The zoo runs a Nest Monitoring Program where volunteers keep an eye on condors in the wild.

COLLABORATING FOR A CAUSE

The San Diego Zoo has a long history with the California condor. In 1929, a condor with a crippled wing was found and taken to the zoo. The wing had to be removed, but the condor lived in the zoo for another 10 years. In 1983, the San Diego Zoo started its Condor Conservation Program by collecting wild eggs to hatch in captivity. Sisquoc was the first to hatch that year. In 1988, Molloko was the first zoo-bred condor to hatch. Since then, 165 hatchlings have been raised at the zoo. Sisquoc is the star of the San Diego Zoo Condor Cam.

populations in the wild—one in California and the other in Arizona. Each population needed 150 birds and at least 15 breeding pairs. The **captive breeding** program soon involved zoos all around the country. The first step was to increase the captive population. This took time, because California condors only have one chick every two years. As the captive condor population grew, the next step was to reintroduce them to the wild.

These condors are ready for release into the wild in Arizona. As well as building up the population numbers, the recovery program also included plans to address the problems of lead poisoning and electrical power lines.

NAMING AND TAGGING

The California Condor Recovery Program tags and numbers all the condors in captivity and in the wild so that individuals can be monitored. Each condor hatchling is assigned a **studbook** number. This helps to keep a record of the bird's medical history, behavior, breeding mates, and offspring. All condors have a tag on their wing with the number. Condors born in the wild are captured briefly to be tagged. The numbers are issued according to age. The higher the number, the younger the bird. The tag number is usually the last one or two digits of the studbook number. The color of the tag identifies numbers in 200s, 300s or 400s.

Visitors can see condors in San Diego Zoo's Condor Ridge enclosures. These condors have been raised at the "condor-minium" in the Wild Animal Safari Park. The condor-minium is a protected site that was built for the condor captive breeding program.

Captive Breeding

The captive breeding program began slowly. **Zoologists** started by collecting eggs from the wild and using **incubators** to hatch them. Once all the condors were captured, scientists started using computer modeling to match male and female birds into breeding pairs. The goal was to ensure **genetic diversity**. All condors alive today are related to just 14 birds.

Most condors mate for life. They stay with the same partner and both parents keep the egg warm and care for the **fledglings** together. It can sometimes take a chick a week to break free from its shell. At about five or six months old, the young condors start learning to fly.

California condors mate for life, so it was important for the captive breeding program to match up unrelated pairs in order to produce healthy chicks with different genes.

DOUBLE CLUTCHING

When captive breeding first began, there was a need to increase the population as quickly as possible. Zoologists used the "double clutching" method. If condors lose a new egg to predators in the wild, they can clutch, or produce another egg, within a month. After a captive condor laid an egg, the zoologists took the egg away to be hatched in an incubator. The breeding pair then laid a second egg, which was hatched by its parents.

PUPPET FEEDING

When birds hatch, they "imprint" on their parents. Imprinting is an automatic response to follow the parent and imitate what it does. In double clutch breeding, many hatchlings had no parents to imprint on. It is important that birds do not imprint on humans because this makes it hard for them to survive in the wild on their own. When the first condors hatched in captivity, scientists put condor puppets on their hands to feed the hatchlings (right). Researchers have used puppets and even full costumes to teach many species of birds and other animals raised in captivity.

Reducing the Risks

Zoos taking part in the Condor Recovery Program were able to treat some California condors for lead poisoning. However, the zoos were overwhelmed by the numbers, because almost every bird feeding on carcasses in the wild needed to have treatment. To protect the condors, **conservationists** worked to reduce the amount of lead in the environment. Hunters of deer and other large mammals were encouraged to replace their lead ammunition with non-lead bullets. This type of hunting helps the California condor population because it increases the safe food supply.

Chris Parish is the condor project director for the Peregrine Fund. In 2015, he and his team captured more than 70 wild California condors in Arizona and Utah. These birds were tested for lead poisoning. Twenty of the birds received treatment and were released back into the wild.

COLLABORATING FOR A CAUSE

In 2008, the state of California passed the Ridley-Tree Condor Preservation Act. The law prohibited lead bullets to be used within the condor's habitat. This law was successful. Lead levels in the condor's blood started to fall. The law also reduced the amount of lead found in golden eagles and turkey vultures. In 2013, the state passed a new law prohibiting the use of lead bullets anywhere in California. The California Department of Fish and Wildlife developed regulations for hunters that went into effect in 2019.

AVOIDING POWER LINES

Power lines are a common problem for wild bird populations. When captive condors were first reintroduced to the wild, collisions and electrocutions were a problem. Scientists developed a plan to train the condors to avoid the power lines before they were released into the wild. The power companies gave electricity poles and wires to zoos involved in the recovery program. Every time a condor landed on a pole or powerline they received a very mild electric shock. The birds soon learned to avoid the power lines. This type of training is called power line aversion training. Similar training was given to other captive birds of prey.

Back from the Brink!

The captive breeding program was a success, and the condor population grew each year. By 1988, zoologists had raised 20 birds. In 1992, the recovery team was ready to start reintroducing condors into the wild. Before releasing captive California condors, the U.S. Fish & Wildlife Service released captive Andean condors into California. Their survival proved that captive condors could survive in the wild. These birds were returned to South America, and eight captive-bred California condors were released in the Sespe Condor Sanctuary, a protected area within Los Padres National Forest in California. But these birds had problems with people and power lines. In 1994, the remaining birds were returned to captivity for their protection. A rethink was needed.

Joseph Brandt, Supervisory Wildlife Biologist for the California Condor Recovery Program, releases a female condor into the wild at the Hopper Mountain National Wildlife refuge. The birds released in 1992 had to be recaptured. They were begging food from people. Some flew into power lines, some died from drinking poison such as antifreeze.

COLLABORATING FOR A CAUSE

In 2014, researchers launched Condor Watch using a website called Zooniverse. This web portal allowed people all over the world to help researchers collect data. Motion cameras took pictures of condors in the wild, but there were too many images for the researchers to review themselves. Volunteers were asked to review 175,000 images of wild condors, noting the tag numbers and the distance between the condor and its food sources. This project collected vast amounts of data, which is being analyzed by researchers. It helps to identify birds at greatest risk.

STAYING AWAY FROM HUMANS

The first release in 1992 showed that condors had to be trained to avoid not only power lines but also humans. Zoo keepers started to harass the fledglings during feedings and pen cleaning. At first, the condors did not seem to mind, but eventually it started to bother them. The training worked. Most condors released into the wild now stay well away from humans.

The recovery program also started to raise condors with their parents or other adult condors as much as possible, instead of using condor puppets. Some condors are now raised in pens in a wild setting, instead of in zoos. This gets them used to the environment. In 1998, five condors were released into the wild without any problems.

Success Stories

Thanks to aversion training, there have only been a few condor deaths due to power lines since 1994. However, as the wild population begins to increase, more condors are hatched and fledged in the wild. These birds do not get powerline aversion training. As young birds, they have less experience and are at a high risk of collision or electrocution on utility poles and power lines. Power companies in California have helped the recovery program by burying power lines and removing the poles in areas inhabited by California condors. Where it has not been possible to bury the lines, the companies have wrapped the wires to avoid the risk of electrocution.

Condors in the wild have radio chips in their tags. Researchers are using these to test an early detection system in wind farms. When a condor flies within 2 miles (3.2 km) of a wind turbine, it shuts down temporarily. The turbine restarts when the condor leaves the area. This should help researchers protect many species of birds in the future.

NO MORE LEAD

The Peregrine Fund partnered with the Arizona Game and Fish Department and Utah Department of Natural Resources. Together they began raising awareness about lead poisoning from ammunition. They educate hunters about non-lead ammunition options such as copper. The hunting community has a tradition of supporting conservation, and most hunters choose to participate in the Condor Recovery Program to protect the birds and the ecosystem.

CALIFORNIA CONDOR RANGE, 2019

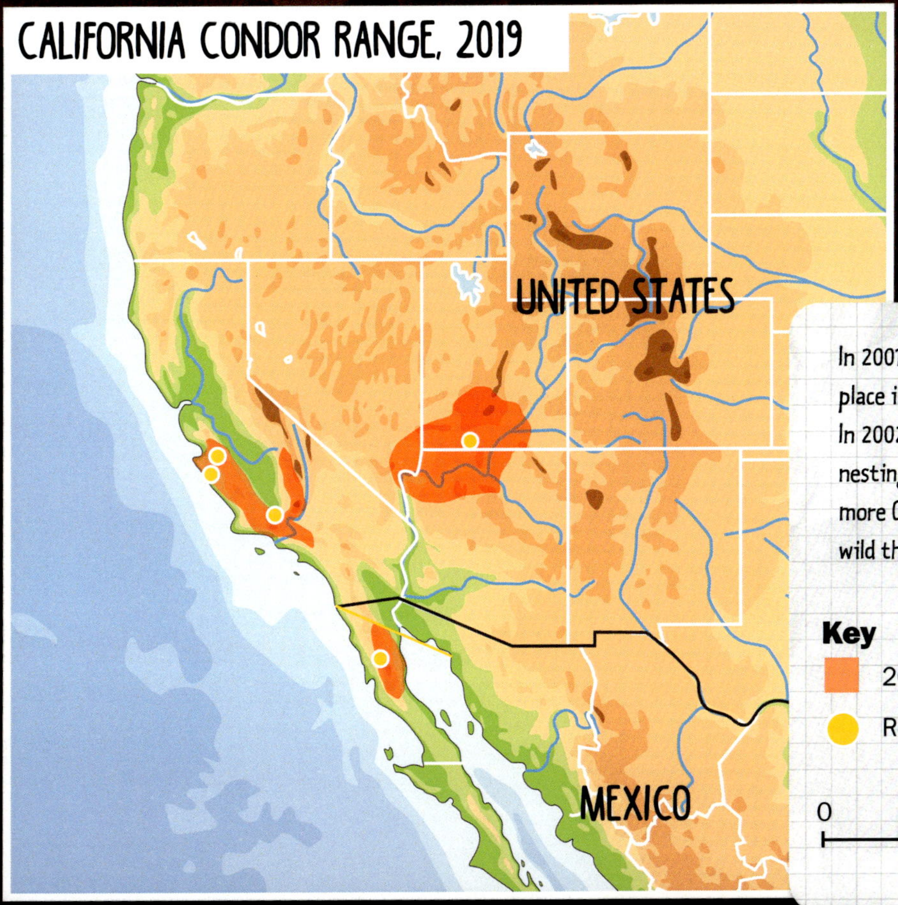

In 2001, the first wild nesting took place in Grand Canyon National Park. In 2002, there were 8 pairs of wild nesting birds. By 2008, there were more California condors living in the wild than in captivity.

What Does the Future Hold?

One important goal for the recovery plan is to spread the California condor population across the entirety of its natural range. To do this, a new release site for captive condors from the United States was found in Sierra de San Pedro Mártir National Park in Baja, Mexico. By 2007, the reintroduced condors had started breeding in Mexico for the first time since the 1930s. Chapultepec Zoo in Mexico City now hatches condors in captivity and releases them into the National Park to increase the California condor population in Baja.

As well as breeding in Mexico, condors are increasing their range elsewhere. In 2014, a condor with tag 597 was seen near San Francisco, much farther north of the central Californian region where it was first released. Condors have also moved north into Utah.

1,000 CHICKS AND COUNTING

In 2019, a special chick hatched in Zion National Park in southwest Utah and took its first flight. This chick was the 1,000th to hatch since the recovery plan started. The hatchling's mother was bred in captivity in San Diego in 2006, and was released into the wild in 2008. The father was hatched in 2009 at the Peregrine Fund's center in Idaho, and was released in 2011. This mother had already hatched two chicks in the wild, but neither of these chicks had fledged, or left the nest. Hatching is no guarantee of survival in the wild.

NEW THREAT

The West Nile virus is a disease spread by mosquitoes (below). It came from northern Africa, near the Nile River. When the virus first appeared in North America in 1999, many bird species were infected and died. There is now a vaccine for birds. In 2005, a wild condor chick died from the virus. The U.S. Fish & Wildlife Service started to vaccinate all captive condors. Since then, every captive bird has been vaccinated. When wild birds are captured for lead poisoning treatment, they are also given the vaccine for West Nile virus. Unfortunately, the vaccine only lasts a short time. The birds have to be revaccinated every few years.

Saving Other Species

As scavengers, condors play an important role in the ecosystem by preventing disease from spreading. Returning the California condor to its natural habitat helps to balance the whole ecosystem and protects other animals and plants.

The California condor recovery effort shows how successful captive breeding programs can be. The sandhill crane was once a bird in danger of becoming extinct in North America. Through captive breeding and habitat protection, its population now numbers around 100,000.

In 2018, the white-backed vulture was listed as Critically Endangered (CR) on the IUCN Red List. The bird was once the most common species of vulture in Africa, but the population is continuing to fall. Loss of habitat, food sources, power lines and hunting have all contributed to the decline in population numbers.

MADAGASCAR FISH-EAGLE

This bird of prey is a medium-sized eagle that lives along the west coast of Madagascar. Its population is about 360 in total, and falling. The species feeds mainly on fish, which competes with the local fishermen. They destroy nests and shoot birds. Habitat loss is also contributing to the fall in numbers. Conservationists aim to increase the numbers through captive breeding programs. In 2018, the IUCN Red List classified the birds as Critically Endangered (CR).

HOODED VULTURE

Like other vultures, the hooded vulture is a scavenger that feeds on the carcasses of dead animals. It lives in sub-Saharan Africa. Threats to the population include poisoning, capture for use in traditional medicine, and hunting for **bushmeat**. As with most other endangered species, loss of habitat is another ongoing threat. As well as land lost to housing developments and other human activity, garbage disposal and hygiene have also reduced food sources. In 2017, the IUCN Red List classified hooded vultures as Critically Endangered (CR).

What Can You Do to Help?

The best way to help protect California condors is to clean up trash. The birds are attracted to microtrash such as scraps and bottle caps, which they take back to their nests with food for their chicks. Microtrash chokes tiny chicks. Arrange regular litter cleanups and reduce your use of single-use plastic. Another way to help is to learn more about the California condor. Share what you find out with family and friends. Social media is a great way to help spread the word!

The U.S. Fish & Wildlife Service offers outreach education programs. Ask your teacher if biologists can come to your school to talk about California condors and how you can help to protect them.

Scientists use web cams to keep an eye on California condor eggs, growing chicks, and their parents. Check out the condor cams run by the organizations mentioned in this book.

TAKE ACTION

You may not live near the California condor's habitat, but there is still plenty you can do to help them and other endangered birds. Here are some ideas you can try:

- Contact your elected officials to tell them that wildlife and habitat protection are important to your future. You can also encourage your federal elected officials to take national conservation seriously. Ask that they approve funds for conservation efforts and pass laws to protect the environment and combat climate change.
- Contact your state and local government to ask what they are doing to keep lead ammunition out of the environment.
- Follow the #californiacondor hashtag on Facebook, Twitter, or Instagram. The Pacific Southwest Region of the U.S. Fish & Wildlife Service (@USFWS_PSW) and the Peregrine Fund (@peregrinefund) also share news about the California condor on social media.

Learning More

Books

Dunn, Mary R. *California Condors.* Capstone Press, 2015.

Lanser, Amanda. *California Condor* (Back from Near Extinction). Core Library, 2016.

Montgomery, Sy. *Condor Comeback* (Scientists in the Field). Houghton Mifflin Harcourt, 2020.

Waxman, Laura Hamilton. *California Condors: Wide-Winged Soaring Birds*. Lerner Publications, 2016.

On the Web

animals.sandiegozoo.org/animals/california-condor
Pages on the endangered California condor from the San Diego Zoo Animals and Plants website. Includes a link to the zoos safari park live condor cam.

defenders.org/wildlife/california-condor
California condor facts, information, and tips on how you can help from the Defenders of Wildlife website. Defenders of Wildlife are a national U.S. conservation organization dedicated to the protection of North American species and their habitats.

endextinction.exposure.co/helping-hands-for-sisquoc
The story of the first zoo hatching of a California condor, with stunning photographs and a rare video of a newly hatched condor chick with its parents.

www.audubon.org/magazine/spring-2018/how-new-technology-making-wind-farms-safer-birds
Pages from the website of the National Audubon Society explaining the new technology helping to protect condors and other birds from wind turbines.

www.peregrinefund.org/projects/california-condor
Photographs and information on the efforts by the Peregrine Fund to save the California condor.

Glossary

breeding pairs A female and male of an animal that mate and have offspring

bushmeat Meat from wild animals, rather than from livestock such as cattle, sheep, and chickens

captive breeding The breeding of animals in captivity by scientists

carcass The body of a dead animal

carrion Decaying flesh of dead or dying animals

conservation Preserving and using resources wisely

conservationist A person who works to preserve resources

ecosystem All living things in a particular area and how they interact

electrocuted To be killed by electric shock from a power source

extinct Describes a situation in which all members of a species have died, so the species no longer exists

fledglings Birds that have flight feathers or that have just left the nest

genetic diversity The amount of different genes in the population of a species

habitat The conditions in which an animal or plant naturally lives

incubators An enclosed box that provides a warm, controlled environment for eggs or young animals

nestling A bird that is too young to leave the nest

New World vultures Scavenger birds found in the Americas, and different from vulture species found in Asia, Africa, and Europe

regurgitation The act of bringing swallowed food back up into the mouth

scavengers Animals that eat carrion, or the flesh of already dead animals

species A group of animals or plants that are similar to each other

studbook A book or database that records information about an animal's parents, grandparents, or great-grandparents.

zoologists Scientists who work in the field of zoology, or the study of animal life

Index and About the Author

ABOUT THE AUTHOR

Rachel Stuckey is a writer and editor with more than 15 years of experience in educational publishing. She has written 30 books for young readers on topics ranging from science to sports, and works with subject experts to develop educational resources in both the sciences and humanities. Rachel travels for half the year, working on projects while exploring the world and learning about our global community.